ACTIVE SCIENCE

Making Sounds

Julian Rowe
and Molly Perham

W
FRANKLIN WATTS
LONDON • SYDNEY

Contents

ACTIVE SCIENCE

Making Sounds

This edition 2004

First published in 1993 by
Franklin Watts
96 Leonard Street
London EC2A 4XD

Franklin Watts Australia
45-51 Huntley Street
Alexandria NSW 2015

Copyright © 1993 Franklin Watts

Editorial planning: Serpentine Editorial
Scientific consultant: Dr. J.J.M.Rowe

Designed by the R & B Partnership
Illustrator: David Anstey
Photography: Peter Millard

Additional photographs:
Chris Fairclough Colour Library 15, 18;
ZEFA 27; The Hutchison Library 30 (top);
Eye Ubiquitous 30 (bottom);
Martin Wendler/NHPA 31.

A CIP catalogue record for this book is
available from the British Library

ISBN 0 7496 5619 0

Printed in Malasyia

 SAFETY WARNING

Activities marked with this symbol require the presence and help of an adult. Plastic should always be used instead of glass.

Listening

Your ears can hear the tiniest sounds.

Whisper quietly in a friend's ear.
Every word can be heard if
you cup your hand.

If it is quiet you
can hear

a
pin
drop.

It helps if you drop the
pin onto a metal tray!
Why do you think this works better?

Clashing cymbals together makes a lot of
noise. You can hear cymbals playing in a
band a long way off. They are very loud.

Hearing sounds

A megaphone helps the sound of your voice reach a long distance away. Most of the sound goes in the same direction instead of spreading out into the air. Directing all the sound together makes it louder.

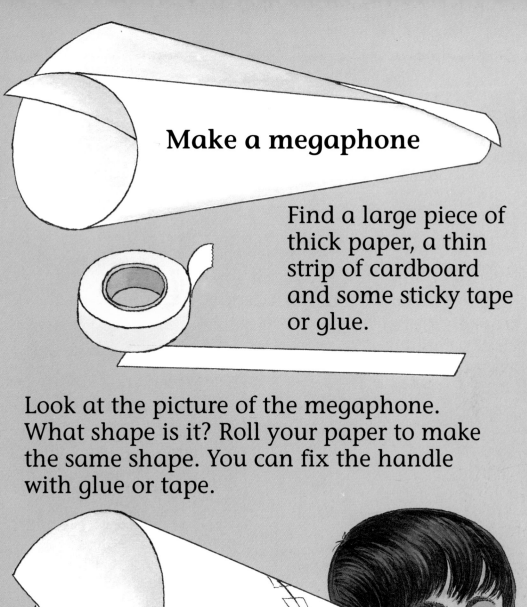

Make a megaphone

Find a large piece of thick paper, a thin strip of cardboard and some sticky tape or glue.

Look at the picture of the megaphone. What shape is it? Roll your paper to make the same shape. You can fix the handle with glue or tape.

Hold your megaphone close to your ear. How well can you hear now?

Noise

Sometimes we wish we did not hear so clearly. Noises are loud sounds we do not want to hear. Very loud noises can damage your ears. Putting your hands over your ears makes the noise softer. People who work in noisy places often wear earmuffs.

This alarm clock has a very loud tick.
Wrapping a wool jumper around the clock
stops some of the sound waves from
escaping.

What do you think it sounds like now?

Travelling sounds

A sound that comes from a long way off takes some time to reach us. We can see something happen before we hear it.

This girl sees the boy banging the cymbals before she hears the clash.

Sound travels through other materials as
well as through air. It travels through
wood. In this picture a watch is taped onto
one side of a door. The girl is standing close
to the door. Do you think she can hear the
watch ticking?

Making a sound

A sound is made by tiny, fast movements of the air. These pass into your ear and make your eardrum vibrate. Messages about the vibrations go to your brain, and so you hear a sound.

When you clap, the air around your hands vibrates and forms waves.

Drop a pebble in some water.

See how the ripples or waves travel outwards in larger and larger circles. The pebble has stirred up the water around it and made the ripples. These spread out like sound waves.

A sound is waves of air. You can hear a sound but you cannot see it.

Speaking

Can you feel a sound? Put your fingers on your throat and say something.

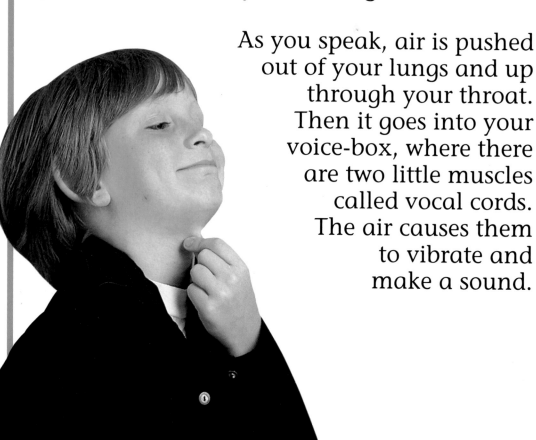

As you speak, air is pushed out of your lungs and up through your throat. Then it goes into your voice-box, where there are two little muscles called vocal cords. The air causes them to vibrate and make a sound.

You can feel these vibrations with your fingers.

Blow up a balloon and stretch the neck out sideways as you let out the air. As you stretch the neck in and out, the sound will change. Your vocal cords work in the same way.

Musical sounds

A musical instrument produces a sound by making the air around it vibrate.

The strings of stringed instruments make the air vibrate when they are plucked or played with a bow.

Wind instruments work when you blow on the column of air inside them.

Percussion instruments have a tight piece of skin or a piece of metal or wood. These make the air vibrate when you bang, scrape or shake them.

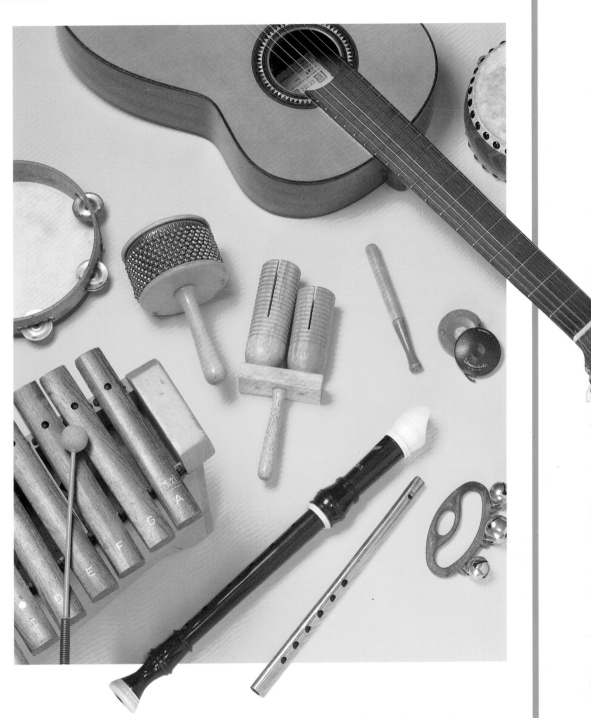

How do these instruments make the air vibrate and so make a sound?

Musical strings

This guitar player plucks the strings to make music. When a string is plucked its vibrations cause sound waves in the air. As these waves reach our ears we hear a note.

The player makes high and low notes by pressing strings with the fingers of his left hand.

Make a guitar

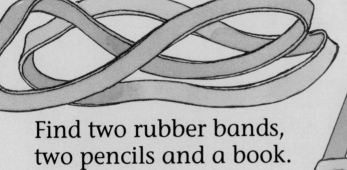

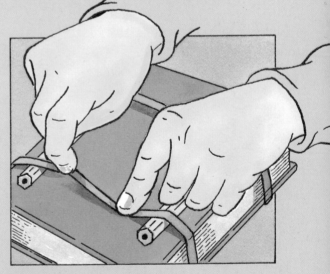

Find two rubber bands, two pencils and a book.

Stretch the bands round the book.

Push the pencils under the bands.

Press down one of the bands with your finger to change the length of the band that you pluck.

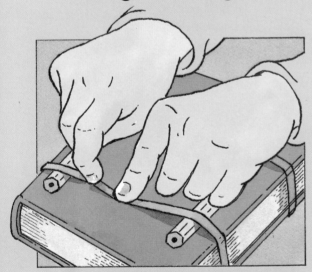

Plucking a short string makes a high note.

Plucking a long string makes a low note.

Musical pipes

The recorder player blows the air inside the pipe to make music. She moves her fingers on and off the holes to produce different notes. When a hole is open near the top of the recorder, some of the air in the pipe escapes. When there is less air it vibrates quickly and produces a high note.

Make a bottle organ

Find six plastic bottles that are the same size and shape. Stand the bottles in a line and pour different amounts of water into each one. Blow across the tops of the bottles. Each one makes a different note.

Which one makes the highest note?

Which one makes the lowest note?

Musical drums

When you bang a drum, its skin vibrates and makes sound waves in the air. This drummer is turning a screw to make the skin tighter.

When the skin is tight, it vibrates faster and produces a high note. When the skin is looser, it vibrates more slowly and produces a low note.

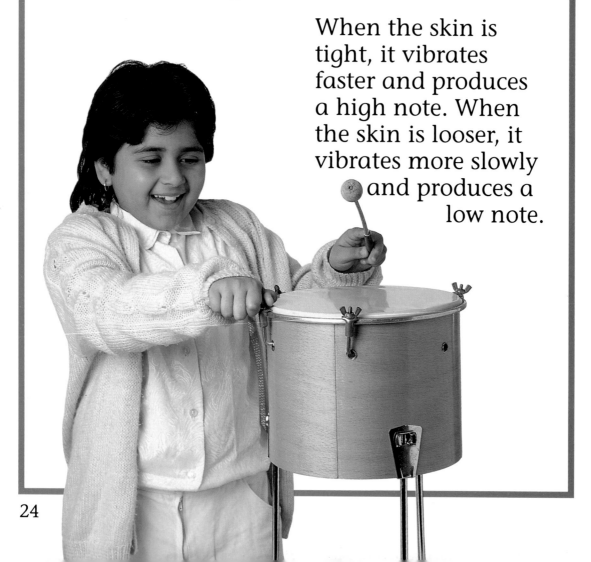

Find a hollow container, plastic wrap or paper, sticky tape or a strong rubber band, and a wooden spoon.

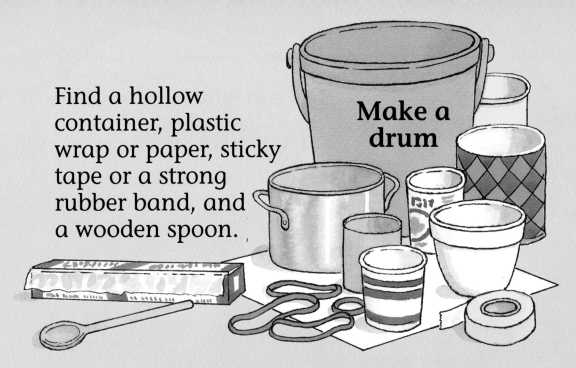

Make a drum

Here are some of the things that you might use. Stretch the plastic or paper tightly over a container to make the drum skin. Tape down the edges firmly, or use the rubber band.

Use the spoon for a drumstick.

Warning sounds

Loud sounds can be useful because they warn us of danger.

A bicycle horn warns people to keep out of the way.

When it is very foggy at sea the lighthouse uses its fog horn to warn ships that land is near.

Can you think of some more warning sounds?

Useful sounds

A whistle can be a warning sound. During a football match the referee blows a whistle to warn players if they are doing something wrong. The whistle also gives information about the time and the rules of the game.

The everyday sounds
around us give us all
kinds of information.
The ringing of an
alarm clock tells us
when it is time to get
up.

When the telephone
rings we know that
someone wants to
speak to us.
A telephone allows
people a very long
way apart to speak
and listen to each
other.

Two-way radios help
firefighters and
police officers
organise help and
rescue people
quickly.

Think about... sounds

This boy cannot hear.
He is deaf.
But he can speak
to other people using
a sign language.

This worker must
protect his ears. The
noise of the road drill
is very loud. Without
the earmuffs, he
would become deaf.

Place the watch at one end of the table. Walk alongside the table until you can't hear it tick. Then place your ear on the table. Can you hear it tick now? Sound travels better in wood than in air.

The fennec fox has huge ears. It uses them to hear the tiny sounds made by the small animals it hunts. What other animals have large ears?